NEXTGEN LEARNERS PRESENTS

Think Like a Brain!™ Collection

Callie the Connector & The Brain Bridge

The Superhighway That Connects Your Thinking Power

BOOK 10

Think Like a Brain!™
Collection Overview

Think Like a Brain!™ is a wildly engaging, silly-smart children's book collection that invites kids to explore how their brains work from the inside out. Through playful characters, imaginative settings, and rhythm-rich storytelling, this collection teaches neuroscience, social-emotional learning (SEL), and self-understanding in ways children absorb without even realizing they are learning.

Instead of telling children how to behave or what to feel, Think Like a Brain!™ shows them why their brains respond the way they do. Each story transforms complex brain science into friendly characters and relatable moments, helping kids feel capable, curious, and empowered rather than confused or ashamed.

Designed for children first and embraced by parents and educators as preventative education, this collection supports:

- Emotional regulation and self-awareness
- Executive functioning skills
- Neurodivergent understanding and inclusion
- Brain-body connection and growth
- Compassion for self and others

With humor, heart, and factual accuracy at its core, Think Like a Brain!™ helps children build the skills many adults wish they had learned sooner, one playful page at a time.

One of the collection's most powerful features is how the characters become real-life reminders for children long after the story ends. Once kids learn who the characters are and what jobs they do, they can begin to recognize those same processes in themselves. A big feeling might become, "It looks like Amy is trying to tell you something," or a moment of pause might sound like, "What does your Fred think?" By naming experiences through familiar characters, children learn to notice what is happening inside their brains, reflect on it, and move forward more quickly with understanding rather than frustration.

Series 1:
The Brainy Bunch (Brain Anatomy)

Series Overview: Meet the brain's core parts as a friendly team. Kids learn what each part does and how they work together to help thinking, feeling, moving, and learning.

1. **The Brainy Bunch**
   Focus: Brain overview and teamwork
2. **Brainstem Benny & The Automatic Adventure**
   Focus: Breathing, heartbeat, survival basics
3. **Bella Cerebellum's Balance Ballet**
   Focus: Balance, coordination, practice
4. **Olly Occipital & The Lookout Tower**
   Focus: Vision, visual processing
5. **Tammy Temporal & The Soundy Sound Storm**
   Focus: Hearing, language, sound processing
6. **Perry the Parietal & The Body Map Bash**
   Focus: Touch, space, body awareness
7. **Hippocampus Hal & The Memory Maze**
   Focus: Memory, learning, recall
8. **Amy the Amygdala & The Overactive Alarm**
   Focus: Emotional alarm system
9. **Luna Lefty & Roger Righty's Mixed Up Masterpiece**
   Focus: Left/right hemispheres, teamwork
10. **Callie the Connector & The Brain Bridge**
    Focus: Corpus callosum, brain-to-brain communication

Series 2:
Cortex Castle (Frontal Lobe Executive Functions)

Series Overview: A lively castle of executive-function helpers. Kids practice planning, starting, stopping, focusing, flexible thinking, time skills, and self-talk, with Frontal Fred guiding thoughtful choices.

1. **Cortex Castle**
   Focus: Executive function overview
2. **Frontal Fred & The Decision Dilemma**
   Focus: Decision-making, prioritizing
3. **Impulsive Ian & The Button Box**
   Focus: Impulse control, pause
4. **Memory Molly & The Wandering Thoughts**
   Focus: Working memory, holding steps
5. **Focus Frannie & The Great Distraction Chase**
   Focus: Attention, refocusing
6. **Planella & The Checklist Parade**
   Focus: Planning, sequencing
7. **Flexi & The Changey-Changeroo Day**
   Focus: Cognitive flexibility
8. **Self-Talk Sally & The Thoughtful Echoes**
   Focus: Self-talk, emotional regulation
9. **Starter Stacy & Finisher Finn**
   Focus: Task initiation, task completion
10. M**ilo the Minute Keeper & Mae the Moment Seeker**
    Focus: Time awareness, pacing
11. **The Power of Pausing with Penny**
    Focus: Pause skills, regulation

Series 3:
Busy Brain Café (Neurodivergent Affirmations)

Series Overview: A welcoming café where different brain styles are celebrated. These stories affirm neurodivergent experiences while teaching self-understanding, communication, and compassion for others.

1. **Nathan Neuron & The Bouncy Brain**
   Focus: ADHD metaphor, movement, attention
2. **Sara Sprinkle the Sensa & The Super Sensitive Day**
   Focus: Sensory sensitivity, overload tools
3. **Overwhelmed Owen & The Senses Overload**
   Focus: Sensory overload, coping strategies
4. **Interoceptive Ivy & The Inside Signals**
   Focus: Interoception, body signals
5. **Zane's Zigzag Thoughts & The Path to Peace**
   Focus: Anxiety, grounding
6. **The Busy Brain Café**
   Focus: Neurodivergent inclusion, different processing styles
7. **The Listening Light**
   Focus: Nonverbal communication, understanding
8. **The Language of Love**
   Focus: Connection, communication, compassion
9. **Trying Theo & The How-To Hiccup**
   Focus: Persistence, frustration tolerance
10. **Drowsy Drew & The Nighttime Repair**
    Focus: Sleep, brain repair
11. **Burned-Out Bonnie & The Rebuild**
    Focus: Burnout, rest, recovery

Series 4:
Steady Systems Squad (Regulation & Growth)

Series Overview: A squad that explains how brains regulate, reset, rest, and grow stronger over time, including stress waves, sleep crews, sensory filtering, and rebuilding after hard days.

1. **The Steady Systems Squad**
   Focus: Regulation systems overview
2. **Timothy Thalamus & The Traffic Tower**
   Focus: Sensory filtering, attention routing
3. **Danny Dimmer & Wake-Up Wonda**
   Focus: Arousal and alertness
4. **The Start, Stop, and Switch Station**
   Focus: Initiation, inhibition, shifting
5. **The Night Shift & The Sleepy Crew**
   Focus: Sleep systems, restoration
6. **The Stress Wave & The Return to Calm**
   Focus: Stress response, recovery
7. **The Brain Growth Gathering**
   Focus: Brain growth and development
8. **Paul's Practice Path**
   Focus: Neuroplasticity
9. **Daisy's Daydream Network**
   Focus: Daydream network, imagination
10. **Remy's Rebuild Crew**
    Focus: Recovery and healing

NEXTGEN LEARNERS PRESENTS

Think Like a Brain!™ Collection

Callie the Connector & The Brain Bridge

The Superhighway That Connects Your Thinking

BOOK 10

Power

WRITTEN &
ILLUSTRATED BY:

MAISEL MCLAULA

NEXTGEN LEARNERS PRESS™
CINCINNATI, OHIO

This book is published by NextGen Learners Press, an imprint of NextGen Learners LLC.

Maisel McLaula is a pen name used by the author of this work.

First Edition: 2026
Printed in the United States of America
ISBN: 978-1-972860-09-0

This book is intended for educational and informational purposes only. It is not intended to replace professional medical, psychological, or therapeutic advice, diagnosis, or treatment.
Always seek the advice of a qualified professional with any questions regarding a medical or mental health condition.

Artwork and visual elements in this book were created using a combination of original artwork, with enhanced original direction, digital illustration tools, and AI-assisted creative technologies.

All content, concepts, characters, and final compositions are the intellectual property of the author and publisher.

A Note for My Readers:

Hi friend,
Your brain is incredible. It thinks, feels, remembers, solves, dances, creates, and so much more!
This series is here to help you understand how your brain works, and to cheer you on as you grow your own thinking superpowers.

Every part of your brain has a job.
Sometimes they get mixed up. That's okay!

When you learn what each part does, you can work with your brain instead of feeling like it's working against you.
So read, giggle, learn, and maybe even draw your own Brainy Bunch when you're done.

I'm cheering you on with every rhyme.

- Maisel McLaula

Think Like a Brain!™

Brain Anatomy Collection

Explore the castle inside your head! Meet each room and its brilliant resident, from Frontal Fred to Benny the Brainstem. Whimsy, wonder, and neurons await!

For Kierslei,
A beautiful reminder that balance and connection make everything stronger.

And...

To the beautifully wired, the curiously wired, and the wonderfully neurodivergent minds, may your messages find their way through.

You are exactly what a great brain looks like.

-Maisel McLaula

Deep in the middle, lives Callie the Connector, tucked out of sight,
The Corpus Callosum who links left and right.
She wasn't a thinker or feeler or seer,
But she made sure both sides could always hear clearer.

She stretched like a bridge from side to side,
With messages zooming along her inside.

She carried ideas both big and small,
Helping the brain work best overall.

Luna Lefty liked logic and numbers and rules,
Checklists and patterns and tidy-up tools.

Roger Righty liked painting and dreaming and play,
Big feelings and pictures and "what ifs" each day.

They both had ideas and worked every day,
But sometimes their thoughts didn't travel the same way.

One had a plan, the other a spark,
But sharing it all could feel tricky and dark.

"Hey, Callie!" said Luna Lefty. "What's Roger Righty up to?"
"I'll check!" said Callie. "And I'll send him you, too!"

Back and forth, she zipped like a beam,
Helping the brain become one thinking team.

When Luna Lefty solved puzzles and lined numbers straight,
Roger Righty imagined and helped ideas create.

Callie passed signals both gentle and fast,
So logic and creativity could happily last.

She helped them take turns and listen with care,
To slow things down or to speed up when aware.

She helped words meet pictures, feelings meet facts,
So thoughts could connect on crisscrossing tracks.

But one busy day, the thoughts got jammed,
Roger Righty felt rushed and Luna Lefty felt crammed.

Ideas piled up and signals crossed,
Plans felt confusing and feelings felt lost.

"Callie, we're stuck!" they cried out in dread.
"We're not on the same page inside our head!"

Callie paused, then breathed real slow,
And gently untangled the traffic flow.

“A team works best when the bridge stays clear,
Not shouting across or frozen in fear.

Slow down your traffic, take one thing at a time,
Let thoughts cross over in rhythm and rhyme."

She sent a calm signal from side to side,
Letting both voices feel heard inside.

Facts met feelings, pictures met plans,
And teamwork returned across brainy lands.

With rhythm restored and cooperation in flow,
The brain hummed smoothly, all systems aglow.

Callie just smiled from her space in between,
The steady connector who keeps thoughts serene.

Tips:

When thoughts feel tangled or you can't decide,
Try checking both Luna and Roger instead of one side.
Ask, "What's the fact? And what's the feel?"
Let both speak clearly, that's how you heal.

You might draw your idea or talk it out loud,
Walk in a circle or lie on the ground.
Helping your brain send signals with care
Can bring your thoughts back into a pair.

What Is the Corpus Callosum?

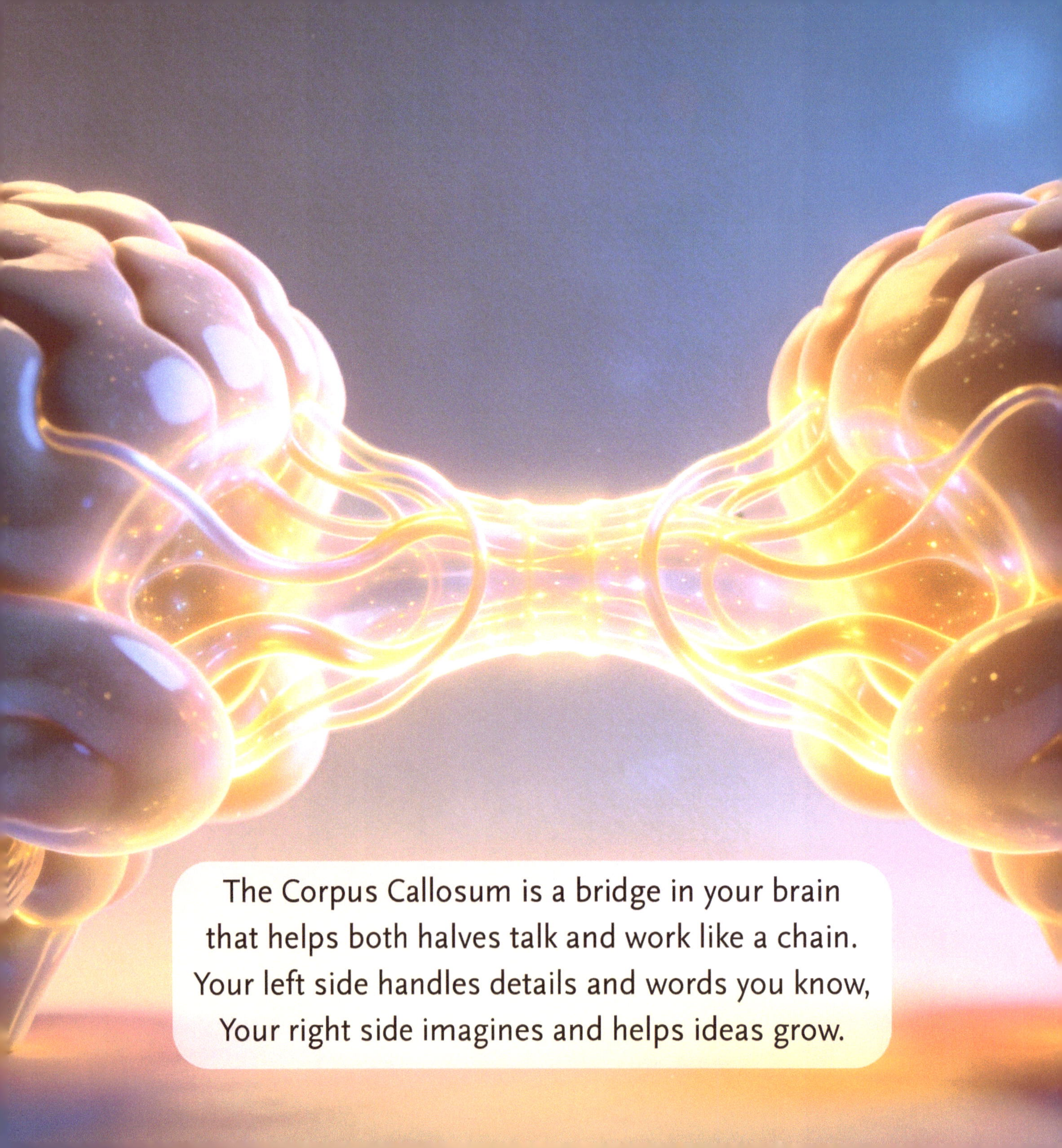

The Corpus Callosum is a bridge in your brain
that helps both halves talk and work like a chain.
Your left side handles details and words you know,
Your right side imagines and helps ideas grow.

Callie connects them so they work as one,
Sharing their strengths so problems get done.
Without her help, thoughts might not link,
Or one side could lead while the other can't think.

Try It Yourself!

Try doing a task that uses both sides,
Like tossing a ball or taking long strides.
Crossing midline with movement or play
Helps Callie's brain bridge clear the way!

You can also draw with both hands in a loop,
Dance side-to-side, or skip in a group.
These crossing moves are a brain-body treat,
Helping your teamwork stay steady and neat.

Congratulations!

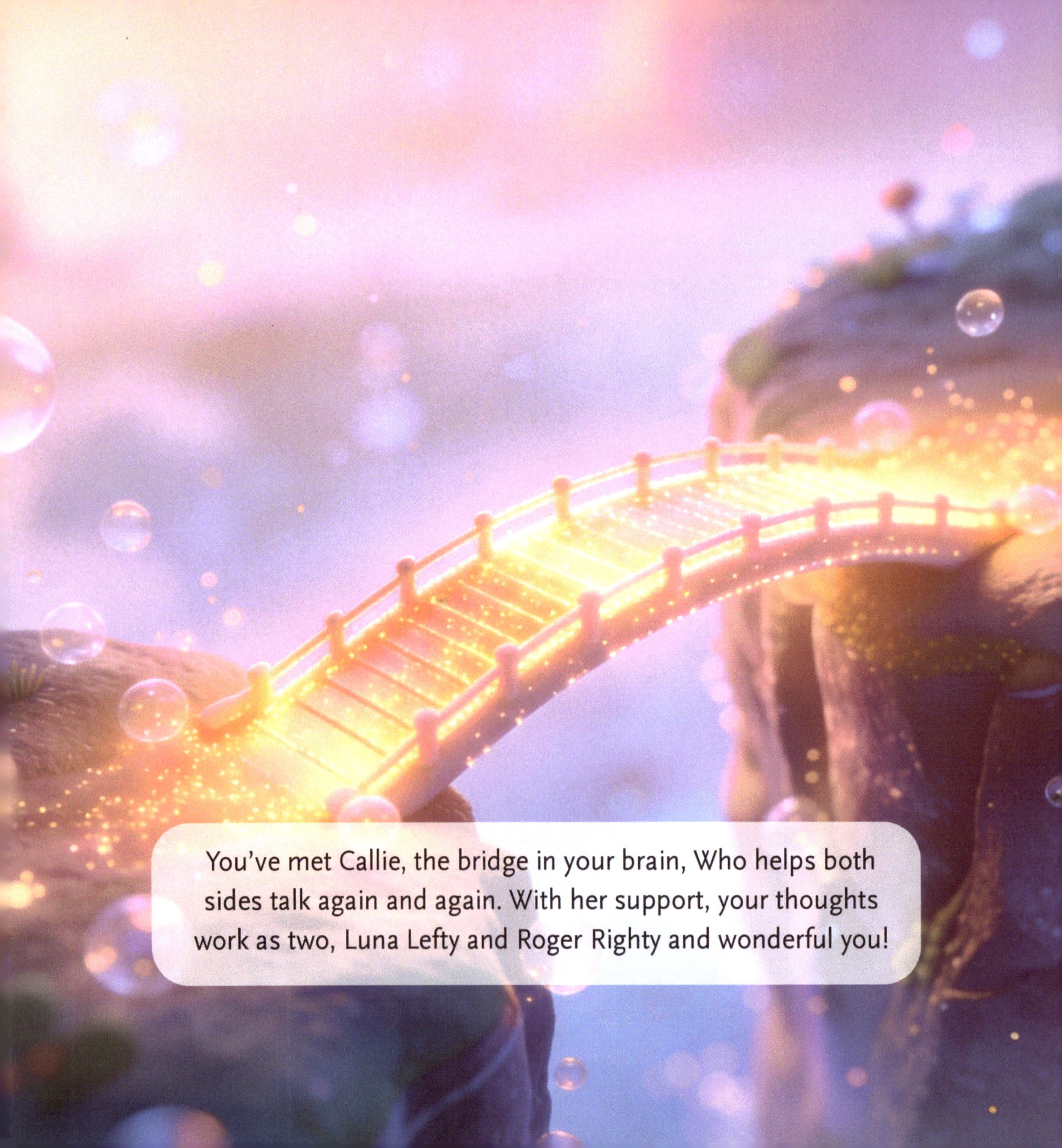

You've met Callie, the bridge in your brain, Who helps both sides talk again and again. With her support, your thoughts work as two, Luna Lefty and Roger Righty and wonderful you!

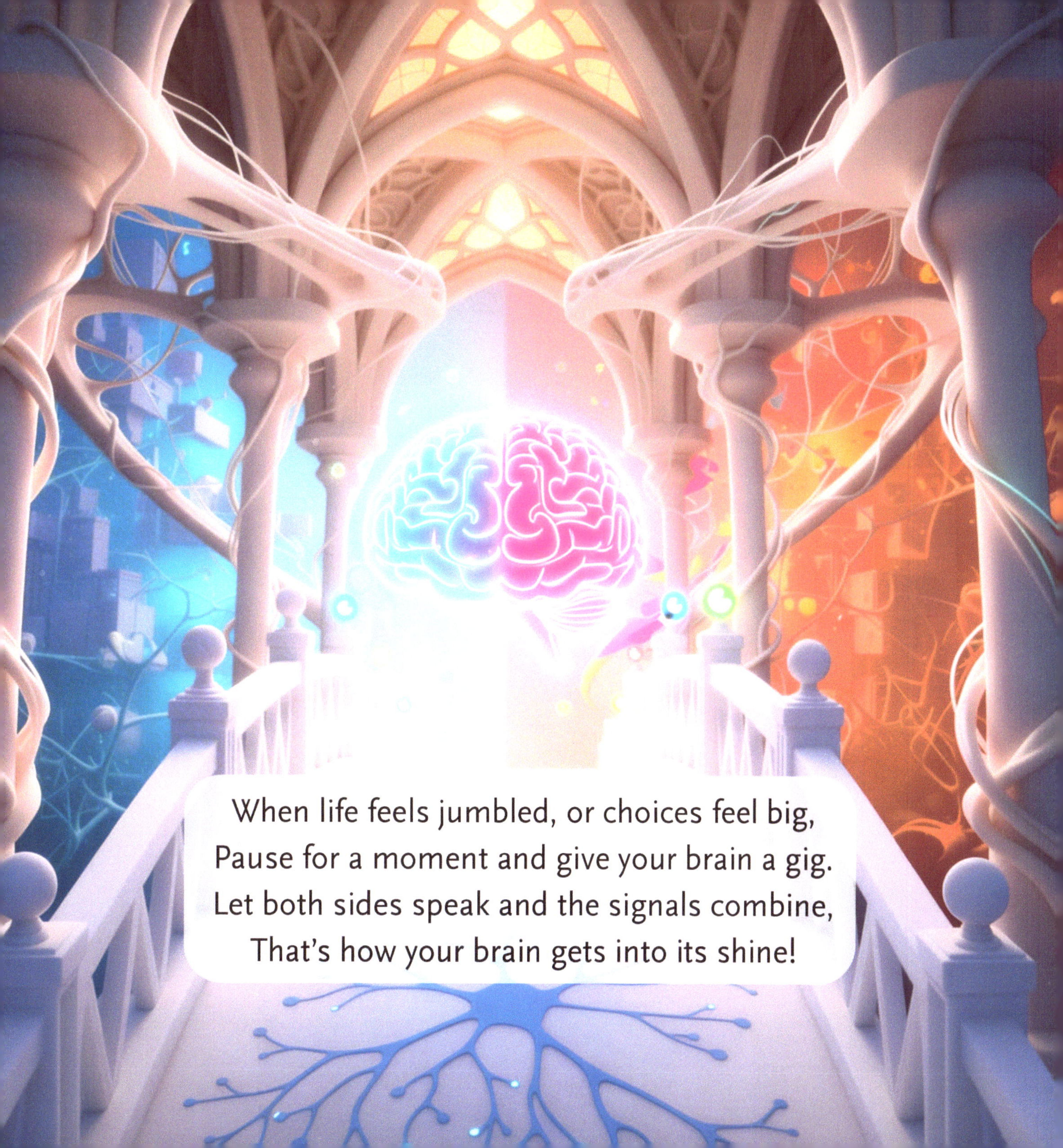

When life feels jumbled, or choices feel big,
Pause for a moment and give your brain a gig.
Let both sides speak and the signals combine,
That's how your brain gets into its shine!

Thank You to My Readers

Thank you for reading this book.
Whether you're a parent, a teacher,
a therapist, or a brilliant young
brain explorer,
your interest in understanding
how the mind works
means the world.

Sharing these books
means spreading compassion and
curiosity.

You are the reason
these stories come alive.
Thank you for believing in the
power of playful learning.

Stay curious. Stay kind.
Stay brainy!

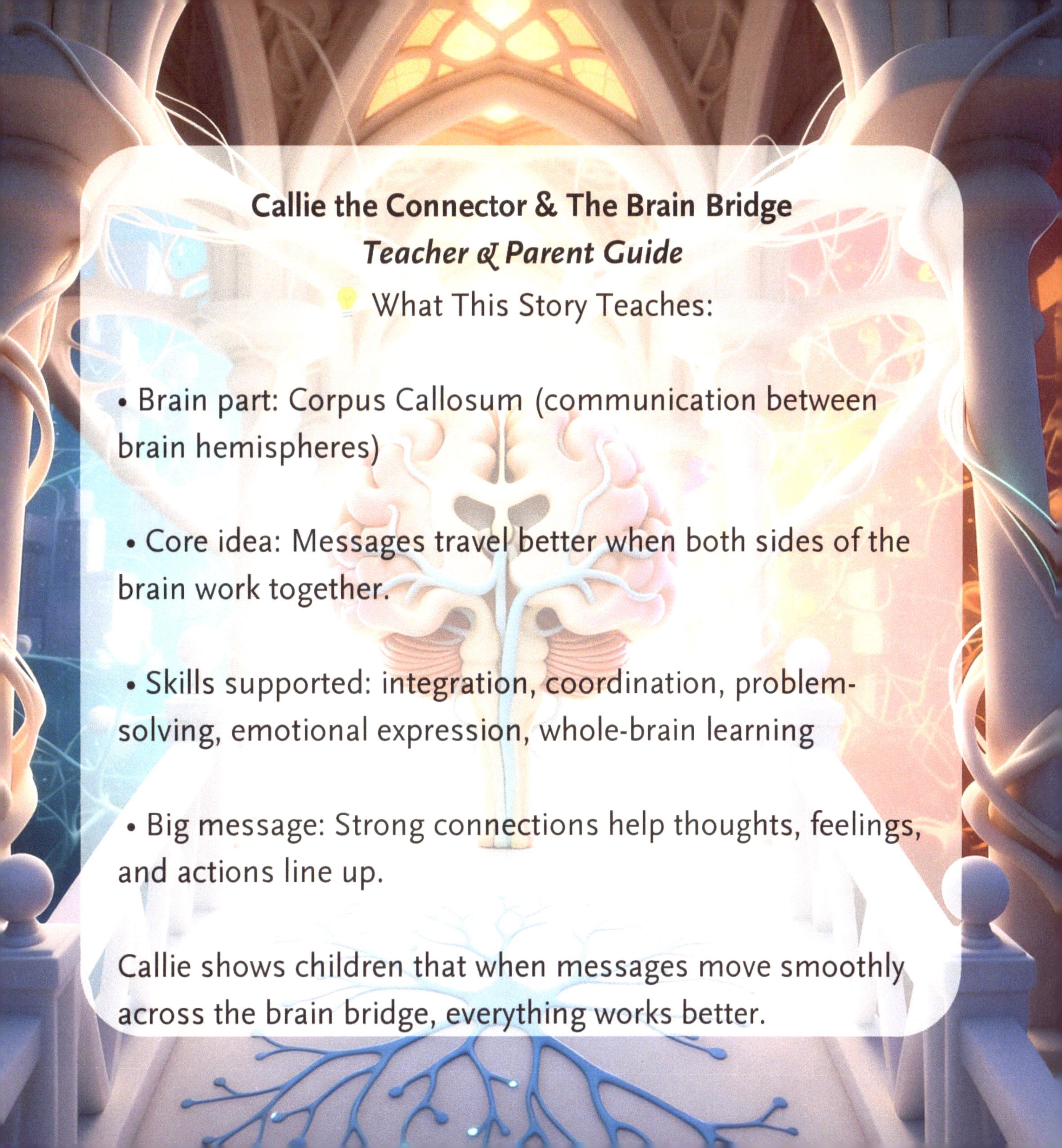

Callie the Connector & The Brain Bridge

Teacher & Parent Guide

💡 What This Story Teaches:

• Brain part: Corpus Callosum (communication between brain hemispheres)

• Core idea: Messages travel better when both sides of the brain work together.

• Skills supported: integration, coordination, problem-solving, emotional expression, whole-brain learning

• Big message: Strong connections help thoughts, feelings, and actions line up.

Callie shows children that when messages move smoothly across the brain bridge, everything works better.

🧠 Key Concepts for Grown-Ups

Callie represents the corpus callosum, a thick bundle of nerve fibers that connects the left and right hemispheres of the brain. This bridge allows information to travel back and forth so logic, creativity, movement, language, and emotion can cooperate.

When communication across this bridge is still developing or under stress, children may:

- Struggle to explain what they feel
- Have trouble coordinating both sides of their body
- Find multi-step tasks difficult
- Get stuck between ideas and actions
- Appear disconnected or scattered

Important foundations:

- Integration develops over time.
- Cross-body movement strengthens the brain bridge.
- Movement plus learning builds communication.
- Emotional safety supports connection.

This story helps children understand that:

- Sometimes messages get stuck.
- Practice helps pathways grow.
- Both sides of the brain need to talk.

🗣️ Discussion Starters

- What happened when Callie's bridge wasn't working well?
- How did Callie help messages move again?
- Have you ever felt like your thoughts and actions didn't match?
- What helps you feel more connected?
- Can you notice both sides of your body right now?

Encourage children to describe experiences without judgment.

👪 Try This Together:

1. Cross-Crawl

Touch right elbow to left knee, then switch. Move slowly and rhythmically.

2. Ball Pass

Pass a ball from one hand to the other across the body.

3. Mirror Play

Stand facing each other and copy slow movements.

4. Two-Hand Art

Draw or color using both hands at once.

5. Walk and Talk

Walk while discussing ideas or practicing spelling words.

🎯 Bonus Challenge

Build Callie's Brain Bridge

Use blocks, paper, or craft supplies to create a bridge.

Ask:

- What helps messages cross?
- What makes bridges stronger?
- What tools help when ideas feel stuck?

Optional Labels:

- Message Lane
- Movement Path
- Feeling Channel
- Thinking Track

Playful Extension:

Bridge Builder Game

Create stations:

- Toss a beanbag across the body
- Crawl under a chair
- Draw a shape with both hands
- Take five cross-crawls

Rotate through each to strengthen integration.

🛠️ Callie's Connector Toolbox:

Use these tools when children seem disconnected or uncoordinated:

- Encourage daily cross-body movement
- Pair learning with motion
- Use rhythm and repetition
- Break tasks into parts
- Support emotional expression
- Slow transitions
- Celebrate effort

Helpful phrases:

- "Let's help your messages travel."
- "Your brain bridge is practicing."
- "Both sides are teaming up."

Avoid rushing integration. Connection grows through gentle repetition.

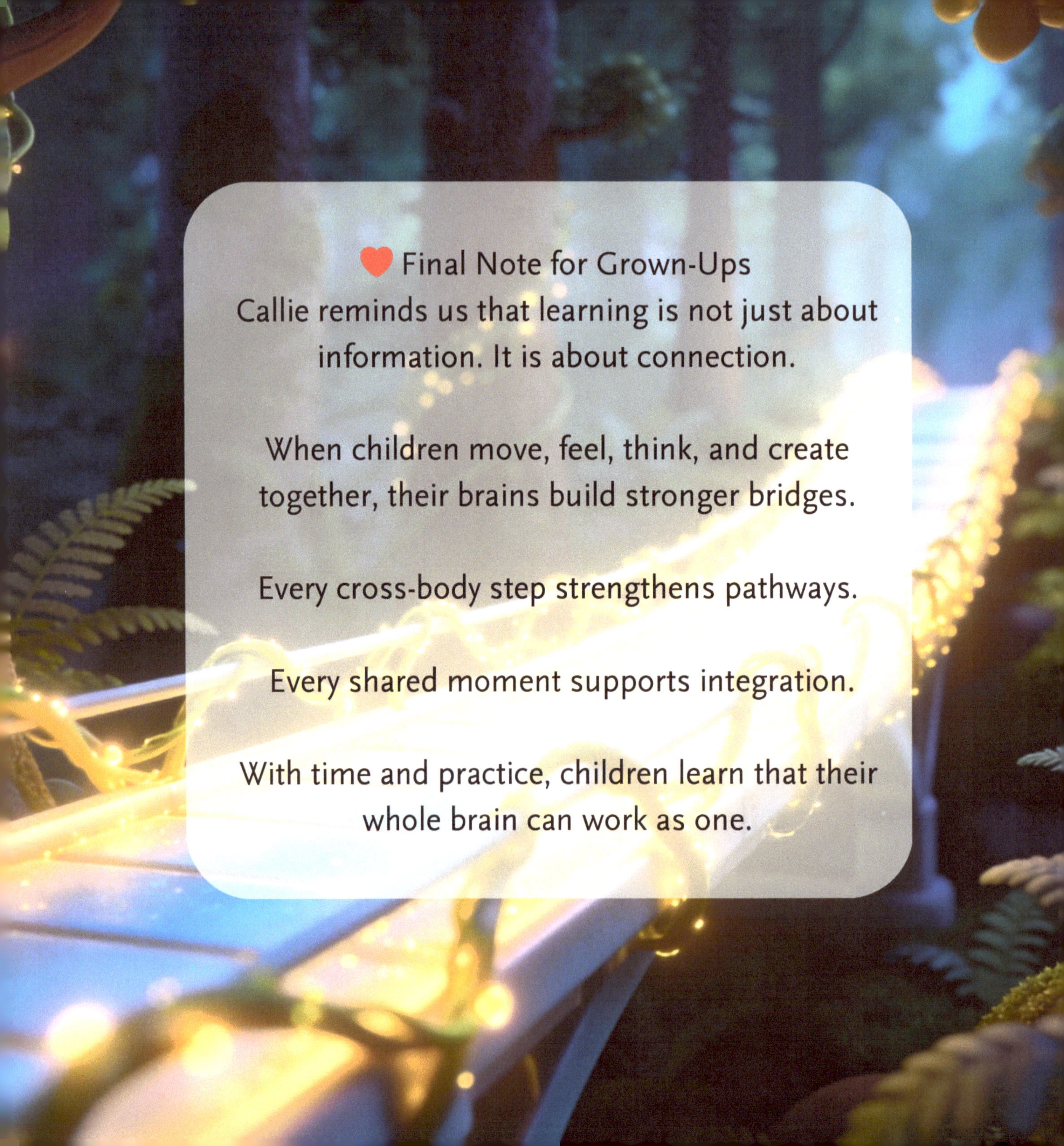

Final Note for Grown-Ups

Callie reminds us that learning is not just about information. It is about connection.

When children move, feel, think, and create together, their brains build stronger bridges.

Every cross-body step strengthens pathways.

Every shared moment supports integration.

With time and practice, children learn that their whole brain can work as one.

Resources &
GROWTH

Resources:

(Corpus callosum + communication between the brain's two sides)

- NINDS (NIH). Brain Basics: Know Your Brain (corpus callosum / brain connections overview).
- Cleveland Clinic. Brain: parts and function (connections/communication between regions).
- NINDS (NIH). Brain Basics: Know Your Brain / Brain Basics resources.
- Cleveland Clinic. Brain: parts and function (overview pages).
- CASEL. CASEL's SEL Framework (self-awareness, self-management, social awareness, relationship skills, responsible decision-making).
- CDC. Child Development: Positive Parenting Tips and developmental guidance by age. Cleveland Clinic
- MedlinePlus (U.S. National Library of Medicine)
- Encyclopedia Britannica

Growth:

(If you or someone you know wants to support integration, coordination, or communication)

- Understood.org
    - Search: coordination skills
- CASEL
    - Look for: relationship skills

Acknowledgments:

To my students: past, present, and future:
You taught me more than any textbook ever could. By watching how you explored, questioned, struggled, and discovered, I learned how minds truly grow. Every "aha" moment in these pages began with you.

To the parents, teachers, and guides walking beside children every day:
Thank you for listening, for advocating when things feel complicated, and for believing that understanding a child's inner world truly matters. Your patience and care shape futures in ways you may never fully see.

To my husband, my best friend, and my constant partner in learning:
Thank you for walking beside me, believing in this work, and holding space for creativity to exist. Sharing curiosity, growth, and creation with you has been one of the greatest honors of my life.

To the great thinkers who shaped how I see the world: including Dr. Seuss, Albert Einstein, Leonardo da Vinci, Stephen Hawking, Ludwig van Beethoven, just to name a few, and my English teacher, Mrs. Godfrey:
Your curiosity, courage, and imagination reminded me that learning is limitless, creativity is powerful, and questions matter. Mrs. Godfrey, your encouragement, belief in me, and love of words sparked my writing journey. If not for you, I may never have picked up a pen.

And to my publisher, NextGen Learners™:
It was by fate that we met, and by inspiration that we now move forward. Together, we build something meaningful. Together, we leave this legacy.

Lastly, to my own brain:
Complex, resilient, ever-adapting, and endlessly learning. Thank you for carrying stories, holding wonder, and reminding me that growth is always possible.

Maisel McLaula is the author of the Think Like a Brain!™ children's book collection, a story-first series that blends neuroscience, imagination, and emotional safety for young readers.

Drawing from decades of experience supporting children and families, her work centers on a simple but powerful belief: children deserve to understand and accept who they are, including how their brains work and why they make the choices they do.

Through playful characters and gentle metaphors, Maisel's stories help children discover the "why" behind their thoughts, feelings, and actions, turning confusion into clarity and self-judgment into self-compassion. By understanding their inner systems, children gain the power to respond with kindness toward themselves and to recognize those same patterns in others.

Maisel believes that when children feel safe inside their own minds, empathy grows naturally. The Think Like a Brain!™ collection offers a foundation for confidence, understanding, and care that grows with the child.

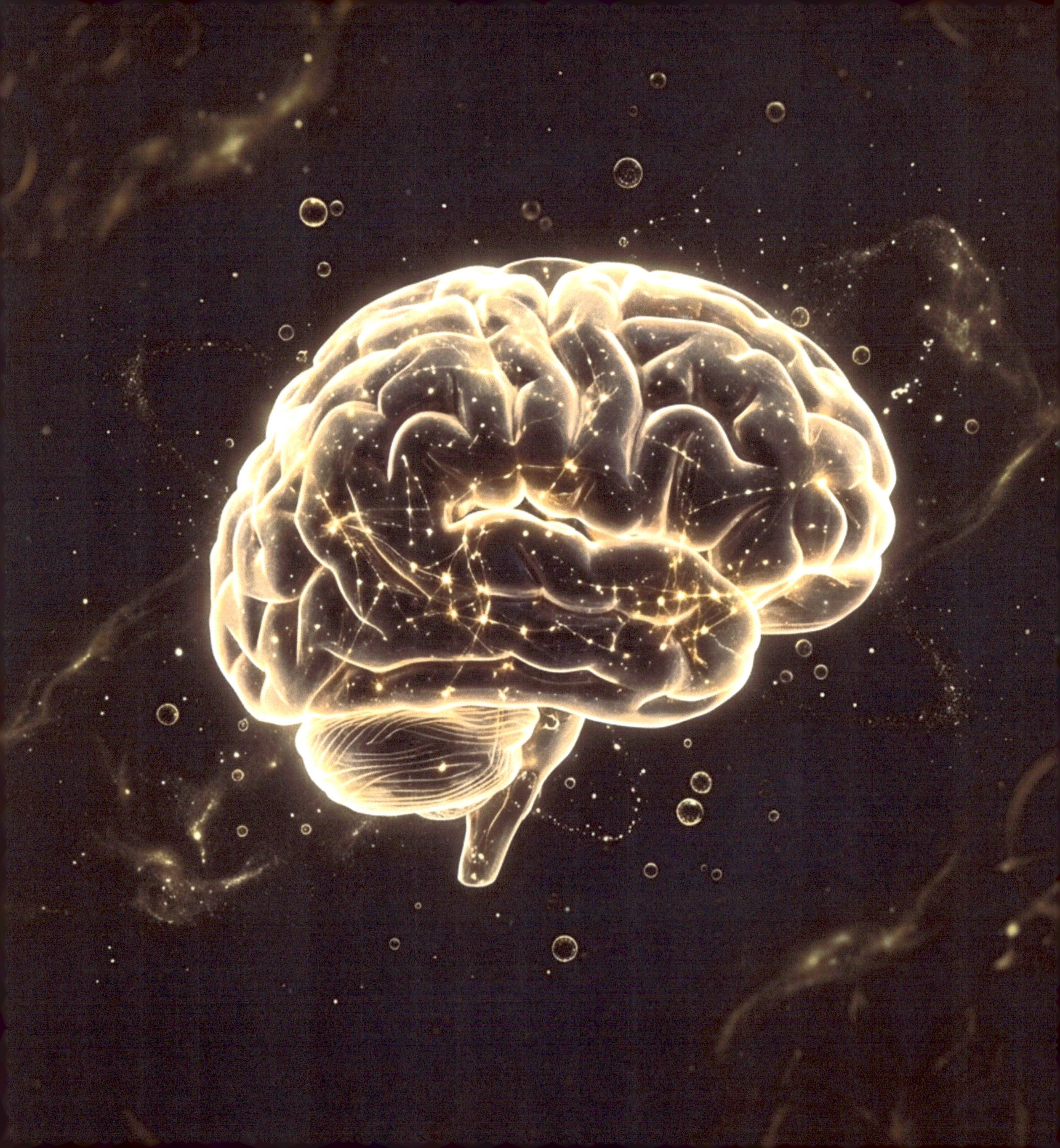

ABOUT THE PUBLISHER

NextGen Learners LLC is an educational publishing company dedicated to helping children understand how they think, feel, learn, and grow.

Through imaginative storytelling, neuroscience-informed concepts, and real-world application, NextGen Learners™ creates books and resources that support emotional development, critical thinking, self-awareness, and confidence in young minds.

Our work is designed not only for children, but for the parents, educators, caregivers, and professionals who walk beside them. Each book serves as a bridge, helping adults better understand the child in front of them, while helping children better understand themselves.

By creating shared language and meaningful connection points, our stories foster deeper relationships, reduce misunderstanding, and encourage acceptance, both within oneself and toward others.

We believe every child's mind is uniquely wired, and that understanding those differences is the foundation for compassion, growth, and lifelong learning.
With multiple collections designed to support cognitive, emotional, and developmental growth, NextGen Learners provides tools that children can carry with them, and tools that adults can use to guide them with clarity, patience, and purpose.

NG
NEXTGEN
LEARNERS
PRESS

MORE TO EXPLORE

The Brainy Bunch is just the beginning.
More stories, tools, and brain-friendly resources are on the way! Designed to help children, parents, and educators better understand how the brain works in everyday life.

FREE PRINTABLES

We've created printable pages to go along with this book, including:

-Reading Guides

-Brain helper visuals

-Character Cards

These are available for free as part of our growing resource library.

Our full website experience is built to include:

-Downloadable activity packs

-Parent & teacher guides

-Expanded learning tools

-Additional collections and topics

STAY CONNECTED

We love hearing how children connect with these stories!

If there is a topic, feeling, or challenge you would like to see explored in future books, we would love to hear from you!

Contact us: Inspire@NextGenLearners.com
NextGenLearners.com

Your ideas help shape future stories and support the growth of this collection.

This collection grows with you. New resources and stories are added regularly.

Maisel McLaula loves hearing from the tiny humans who read her books.

Did a story make you giggle, spark a big idea, or inspire you to draw your very own character? She would be delighted to see it! Send your letters and artwork to:

NextGen Learners
Attn: Maisel McLaula
8666 Beechmont Ave #1063
Cincinnati, OH 45255
Every letter is read with a smile, and your artwork just might make her whole day.

OPEN BRAIN

www.ingramcontent.com/pod-product-compliance
Lightning Source LLC
Chambersburg PA
CBHW041322210926
51847CB00002B/21
9781972860090